日精進
rì jīng jìn

成杰 著

青少年双语版

Bilingual Edition of Make Progress Day by Day for Teenagers

青少年成长必读的36封双语励志信

中国水利水电出版社
www.waterpub.com.cn
·北京·

内容提要

《日精进·青少年双语版》集成了作者成杰老师10多年来成功和励志的智慧结晶,以语录为主体,从成长、梦想、学习、教育、智慧、幸福等篇章宣扬了日日精进的关键亮点,也契合了《礼记·大学》中"汤之盘铭曰:苟日新,日日新,又日新"的日精进学习精神。本书为中英文双语版,采用裸脊装,四色印刷,配有精美插图,分为36个章节,每一章节都是写给青少年的一封信,希望青少年能够将成长当作人生的头等大事,每天都进步一点点,天天向上,日日精进,早日成功。

图书在版编目(CIP)数据

日精进:青少年双语版:汉英对照 / 成杰著. --
北京:中国水利水电出版社, 2022.5
　ISBN 978-7-5226-0442-8

Ⅰ.①日… Ⅱ.①成… Ⅲ.①成功心理—青少年读物
—汉、英 Ⅳ.①B848.4-49

中国版本图书馆CIP数据核字(2022)第019781号

书　　名	日精进·青少年双语版 RI JING JIN·QINGSHAONIAN SHUANGYU BAN
作　　者	成杰 著
出版发行	中国水利水电出版社 (北京市海淀区玉渊潭南路1号D座 100038) 网址:www.waterpub.com.cn E-mail:zhiboshangshu@163.com 电话:(010)68367658(营销中心)
经　　售	北京科水图书销售中心(零售) 电话:(010)88383994、63202643、68545874 全国各地新华书店和相关出版物销售网点
排　　版	北京智博尚书文化传媒有限公司
印　　刷	北京富博印刷有限公司
规　　格	148mm×210mm　32开本　6.5印张　81千字
版　　次	2022年5月第1版　2022年5月第1次印刷
印　　数	00001—15000册
定　　价	58.00元

凡购买我社图书,如有缺页、倒页、脱页的,本社营销中心负责调换
版权所有·侵权必究

成杰
CHENG JIE

巨海集团董事长
上海巨海成杰公益基金会发起人

成杰智慧心语

日精進 青/少/年/双/语/版

Make Progress Day by Day

序言
Foreword

何为智慧？
What is wisdom?
每日求知为智，
Wisdom comes from a thirst for knowledge every day
内心丰盛为慧。
and the richness of mind.

人的一生中都在努力追求着成长、成功、财富、梦想、付出与贡献，这一切的实现都离不开一个人的智慧。
A person strives for growth, success, wealth, dreams, dedication and contribution throughout his life, which cannot be achieved without his wisdom.

一个人的智慧来自哪里呢？
Where does a person's wisdom come from?
最简单、最直接的方式就是来自学习、成长和精进。
The simplest and most direct way to acquire wisdom is to learn, grow and make progress.

日有所学，

Learn day by day,
月有所获,
you will gain month by month
年有所成。
and make achievements year by year.

每天进步一点点,就是迈向卓越的开始。不积跬步,无以至千里;不积小流,无以成江河。无一日不成长,无一日不精进,势必攀登人生的顶峰。
Making a small progress every day makes the way for excellence start. If there is no one-step and half-step accumulation, there will be no way to reach thousands of miles; If there is no accumulation of small brook, rivers and seas cannot be formed. If a person grows and makes progress every day, he is bound to reach the peak of his life.

《日精进》系列书籍自 2015 年出版以来深受读者朋友的喜欢,很多企业家在读《日精进》一书时也影响着自己的孩子,使其一起日日精进,向上向善,于是大家纷纷建议出版一本《日精进》青少年版。
Since it was published in 2015, the series of *Make Progress Day by Day* have been loved by readers. When reading the

book *Make Progress Day by Day*, many entrepreneurs' children were also affected. They made progress, kept uplifting and good-hearted during the same time. Therefore, many people suggested publishing an edition of *Make Progress Day by Day* for teenagers.

巨海第 50 期"演说智慧"演讲冠军、南部东辰国际学校党支部书记陈胜校长将《日精进》一书运用在学生教育中，短短一年时间取得了惊人的成效，这给了我很大的信心和动力。后来在陈胜校长支持下推动了此书的出版，于是就有了今天大家所阅读到的《日精进》青少年双语版，在此对陈胜校长的付出表示诚挚的感谢！

Chen Sheng, party branch secretary and principal of Nanbu Dongchen International School, the speech champion of the 50th issue of Speech Wisdom of Juhai, applied the book *Make Progress Day by Day* to the education of students and achieved amazing results in just one year ,which gave me great confidence and motivation.After that, the joint support of principal Chen Sheng， promoted the publication of this book, and thus, we have the bilingual edition of *Make Progress Day by Day* for teenagers that you read today. I would like to express my sincere thanks to president Chen Sheng for his efforts!

日日行，不怕千万里；
Travel day by day, and you will not fear thousands of miles;
常常做，不怕千万事。
Do things frequently, and you will not fear most things.

让我们以日日精进之精神来成长自己，
Let's grow with the spirit of making progress day by day,
最终遇见更好的自己！
so that we will eventually meet a better version of ourselves!

成杰
Cheng Jie
2021 年 10 月 10 日于上海
October 10th , 2021 in Shanghai

目录 CONTENTS

青少年演说家·宣言 ………………………… 001
Young Speaker - Manifesto

第1封信：
超越梦想 …………………………………… 008
Letter 1: Beyond the Dream

第2封信：
爱自己最好的方式 ………………………… 013
Letter 2: The Best Way to Love Yourself

第3封信：
　　我的心中有个巨人 019
Letter 3: A Giant in My Heart

第4封信：
　　人生不能不成长 024
Letter 4: You Have to Grow up in Life

第5封信：
　　拥有梦想，就拥有希望 029
Letter 5: Where There is a Dream, There is a Future

第6封信：
　　读书让你拥有智商 034
Letter 6: Study Offers You Intelligence

第7封信：
　　把语言化为行动 041
Letter 7: Put Your Words into a Action

第8封信：
　　梦想的大小，决定成长的快慢 046
Letter 8: A Greater Dream Will Be Followed by a Quicker

日精進 青/少/年/双/语/版

Make Progress Day by Day

Chaser

第9封信：
积极的人在忧患中看到机会 052
Letter 9: Positive People Find Opportunities in Hardship

第10封信：
遇见明天更好的自己 056
Letter 10: Meet a Better You Tomorrow

第11封信：
成功是累积而成的 062
Letter 11: Success Comes from Constant Efforts

第12封信：
成为时代的弄潮儿 068
Letter 12: Being a Trendsetter of the Times

第13封信：
命运并非机遇，而是一种选择 072
Letter 13: Destiny is Not an Opportunity, But a Choice

第14封信：
学习的敌人是自己的满足 ………… 076
Letter 14: Complacency is the Enemy of Study

第15封信：
人生是自我期许的结果 ………… 081
Letter 15: Life is a Product of Self-expectation

第16封信：
教育的核心价值 ………… 087
Letter 16: The Core Value of Education

第17封信：
每次失败都会使人更坚定 ………… 093
Letter 17: Every Failure Makes People Firmer

第18封信：
积极的人控制环境 ………… 098
Letter 18: Positive People Govern the Environment

第19封信：
梦想是人生的导航仪 ………… 104
Letter 19: Dream, The Navigator of Life

第20封信：
　学习获得知识 108
Letter 20: Learning To Acquire Knowledge

第21封信：
　格局一旦放大，美好即将发生 111
Letter 21: A Beautiful Life Comes from Wide Horizon

第22封信：
　与高手同行并成为高手 115
Letter 22: Become a Master-hand Under the Guidance of Master

第23封信：
　学习的最高境界 120
Letter 23: The Highest Realm of Learning

第24封信：
　内心强大则世界辽阔 123
Letter 24: Stronger Heart, Wider World

第25封信：
  没有经历付出就不会杰出 128
Letter 25: No Pains, No Gains

第26封信：
  目标一旦确定，方法就会出现 132
Letter 26: A Clear Goal is Followed by Ways

第27封信：
  开卷有益，开口有才 136
Letter 27: Be Beneficial from Books, Be Talented from Speaking

第28封信：
  志不立，天下无可成之事 142
Letter 28: No Vision, No Accomplishments

第29封信：
  利众者伟业必成 148
Letter 29: Great Achievements are Made for the Public Good

第30封信：
　　地低成海，人低为王 153
Letter 30: The More Noble, the More Humble

第31封信：
　　所有世间乐，悉从利他生 156
Letter 31: All Happiness in the World Comes from Helping Others

第32封信：
　　智慧而淡定，仁爱而从容 162
Letter 32: Wise and Composed, Kind and Calm

第33封信：
　　不辩，是一种智慧 168
Letter 33: It is a Kind of Wisdom not to Argue

第34封信：
　　人生的成长是日精进和随时学 172
Letter 34: Make Progress Every Day and Learn at Any Time for Improvement

第35封信：
平常心是道，道是平常心 178

Letter 35: An Ordinary Mind is Excellent Practice, and Excellent Practice Comes from an Ordinary Mind

第36封信：
生命智慧的十大法门 184

Letter 36: Ten Ways of Life Wisdom

日精進 青/少/年/双/语/版

Make Progress Day by Day

学习是智慧的升华，
分享是生命的伟大。

青少年演说家·宣言

Young Speaker – Manifesto

今天是我新生命的开始，

Today is a fresh start for me,

我立志要成为超级演说家，

I am determined to become a super speaker

用演说改变命运，

to change my destiny with speeches,

日精进 青/少/年/双/语/版

Make Progress Day by Day

用演说塑造生命，

to shape my life with speeches,

用演说创造奇迹，

to create miracles with speeches,

用演说谱写传奇。

and to write legendary stories with speeches,

我要用演说去帮助、影响和成就更多的生命！

I'm going to make speeches to help and influence more lives to attain what they want.

从今天开始，

From today on,

我要坚持阅读，

I'm going to keep on reading,

丰富我的知识，

to get a wealth of knowledge.

因为我坚信：

Because I firmly believe

超级演说家一定是博学广闻的人。

a super speaker must be a knowledgeable one.

从今天开始，

From today on,

我要大量朗读，

I'm going to read extensively,

练习我的演讲，

to practice my speech skills.

因为我坚信：

Because I firmly believe,

超级演说家是不断练习的结果！

continuous practice will help train a super speaker!

日精进 青/少/年/双/语/版

Make Progress Day by Day

从今天开始,

From today on,

我要采取大量的行动,

I'm going to take massive actions,

创造属于自己的故事,

to create my own story.

因为我坚信:

Because I firmly believe,

超级演说家,

a super speaker must,

一定是:做自己所说,说自己所做的人!

do what they promise to do and say what they have done!

从今天开始,

From today on,

我要把演说融入我的生命,

I'm going to make speeches part of my life,

我要日日精进、向上向善,

I'm going to make daily progress and keep uplifting and good-hearted,

持续成长自己，成就他人！

I'm going to continue to improve myself and help others become better!

我要成为：

I will try my best to make myself

同学的榜样，

an example of my classmates,

父母的骄傲，

the pride of my parents,

家族的荣耀，

and also my family,

我言行一致，知行合一，

I will walk the talk and keep the unity of knowledge

and action,

我说到，就一定要做到，

I will act on what I say

没有谁，可以想象我的未来有多么的辉煌！

I believe as long as I try my best, I will have a bright future.

我热爱演讲，

I love making speeches,

我乐于分享，

and I am happy to share,

我就是青少年演说家！

I am a young speaker!

熟记核心价值观，
言行举止有规范。
二十四字记心田，
我为祖国谱新篇。

第 1 封信：

超越梦想

Letter 1: Beyond the Dream

一个不能日日精进的人，

One who cannot improve himself every day

就是在背叛自己的梦想；

will not reach his dreams.

一个不断自我超越的人,

One who keeps outdoing himself

就是在呵护自己的梦想。

will make his dreams come true sooner or later.

心小,则事大;心大,则事小。

Narrow minds lead to big troubles while broad minds make troubles smaller.

念头,是种子;行动,是扎根。

Ideas are seeds, and actions are roots.

只有在泥土中深根百丈,

Only when deeply rooted in the soil

才能在蓝天中际会风云。

can it greet the blue sky.

时光飞逝,唯有实力永存。

As time flies, only strength works forever.

生命如此伟大，拥有无限可能。

Life is so great that it holds infinite possibilities.

人生不设限，才能精彩无限。

Life can be wonderful without limits.

人生没有如果，人生只有结果。

There is no assumption but results in life.

困难像弹簧，你强它就弱，你弱它就强。

As difficulties are like springs, they will loom large if you are weak.

熟能生巧，巧能生精，精能生妙，妙能入道。

Skill comes from practice, and practice makes perfect.

格局大，则虽远亦至；

With large vision, you can arrive in spite of a long distance.

格局小，则虽近亦阻。

With small vision, you will be impeded even at a short ; distance.

只有在泥土中深根百丈，才会在蓝天中际会风云。

日積進 青/少/年/双/语/版
Make Progress Day by Day

穷人因书而富
富人因书而贵

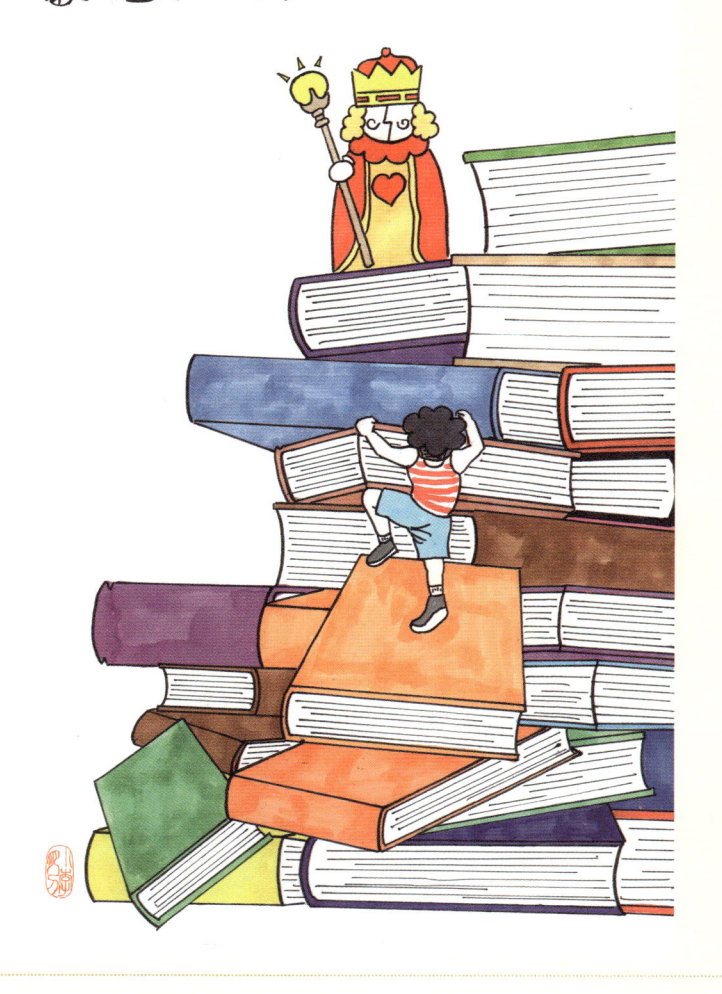

第 2 封信：

爱自己最好的方式

Letter 2: The Best Way to Love Yourself

爱自己最好的方式就是成长自己。

The best way to love yourself is to make yourself better.

把时间用在学习上，把心思用在成长上。

Spend your time learning and improving yourself.

日精進 青/少/年/双/语/版

Make Progress Day by Day

数百年世家无非积德,

A well-known family lasts hundreds of years because of nothing more than virtuous actions,

第一件好事还是读书。

the top virtuous action is study.

读书会让我们遇见更好的自己。

Study will make us better.

什么是成功?

What is success?

成功就是今天比昨天更有智慧,

Success means being wiser today than yesterday,

比昨天更懂得爱,

understanding love better than yesterday,

比昨天更懂得生活的美,

comprehending the beauty of life better than yesterday,

比昨天更懂得宽容别人。

and knowing better how to be tolerant of others than yesterday.

成功的人就是日日精进、向上向善的人!

Those who are successful make progress day by day and keep uplifting and good-hearted.

穷人因书而富,富人因书而贵。

Reading helps the poor achieve wealth, and the rich become noble.

问题就是礼物,问题就是课题。

Problems are gifts and opportunities for learning.

合理安排时间,就等于节约时间。

To well arrange time is to save time.

含泪播种的人，定能含笑收获。

Those who move forward with tears will be generously rewarded.

人生要懂取舍，人生要知进退。

We should know when to hold fast and when to let go.

思考问题的品质，决定了人生的品质。

The quality of thinking determines the quality of life.

学习是智慧的升华，

Learning can make one wiser,

分享是生命的伟大。

and sharing can make life greater.

做热爱并擅长的事，

Doing what you love and do well in

便是最美好的人生。

is how the best life looks like.

含泪播种的人，
必能含笑收获。

日積進 青/少/年/双/语/版

Make Progress Day by Day

我的心中有个巨人，
从今天开始将他唤醒。

第 3 封信

我的心中有个巨人

Letter 3: A Giant in My Heart

我的心中有个巨人，

There is a giant in my heart,

从今天开始将他唤醒。

I'm going to wake him up from today.

我最大的敌人是我自己，

My worst enemy is myself

我最大的救星也是我自己。

and my greatest savior is also myself.

不怕万人阻挡，只怕自己投降。

I am not afraid of being stopped by thousands of people, but surrendering myself.

使我痛苦者，必使我强大。

Those who make me miserable will make me stronger.

人生最大的危机就是：

The biggest crisis in life is

比我更优秀的人，反而比我更努力。

those who are better than me also work harder than me.

生命的蜕变在于真正决定。

A radical change of life is supported by a firm determination.

我愿意做一个有智慧、有胸怀、有格局、

I am ready to become a wise, broad-minded and visionary person

有境界、有追求、有担当、有宽容之心的人。

with thoughts, pursuits, responsibilities and tolerance.

我愿意：

I am ready to

做好自己，把握当下，

be myself and seize the moment,

日日精进，向上向善，

make progress day by day and keep uplifting and good-hearted,

去帮助、影响和成就我所遇到的每一个生命！

help and encourage everyone I meet to attain what they want.

心胸决定格局，

The mind determines vision,

格局决定布局，

the vision determines arrangement,

布局决定结局。

and the arrangement determines outcome.

人生可以不成功，
但人生不能不成长。

第 4 封信：

人生不能不成长

Letter 4: You Have to Grow up in Life

树要成长要有根，

Trees need roots to grow

人要成长要有心。

while people need resolution to grow.

人生可以不成功，

Success is not a must

但人生不能不成长。

but we have to make progress in life.

成长是迈向成功的必经之路。

Growing up is the only way to achieve success.

没有成长的成功，是偶然的成功；

Without growth, success happens by chance.

没有成长的成功，是运气的成功；

Without growth, success happens in luck.

唯有成长之后的成功，才会持续永恒。

Only after experiencing growth will success last forever.

今天，无数的人都在追求成功与名利，

Numerous people are pursuing success, fame and wealth today,

日精進 青/少/年/双/语/版
Make Progress Day by Day

却只有少数的人能静下来，

but only a few can calm down

追求自己的成长与精进。

to pursue their own growth and improvement.

一个人外在的成功和成就，

One's external success and achievements,

是他内在成长和成熟的体现。

are reflective of his internal growth and maturity.

日日行，不怕千万里；

Travel day by day, and you will not fear thousands of miles.

常常做，不怕千万事。

Do things frequently, and you will not fear most things.

让我们立即行动。

Let's take actions right now.

坐而论道,起而践行。

To engage in practice instead of sitting down for discussions.

日日精进,向上向善。

To make progress day by day and keep uplifting and good-hearted.

日精進 青/少/年/双/语/版
Make Progress Day by Day

拥有梦想，就拥有希望；
拥有梦想，就拥有方向；
拥有梦想，就拥有能量；
拥有梦想，就拥有力量；

第 5 封信：

拥有梦想，就拥有希望

Letter 5: Where There Is a Dream, There Is a Future

拥有梦想，就拥有希望；
Where there is a dream, there is a future.
拥有梦想，就拥有方向；
Where there is a dream, there is a direction.

拥有梦想，就拥有能量；

Where there is a dream, there is energy.

拥有梦想，就拥有力量。

Where there is a dream, there is power.

做自己所说，说自己所做。

Do what you promise to do and say what you have done.

言行一致，知行合一。

I will walk the talk and keep the unity of knowledge and action.

简单的事情重复做，就是专家；

Repeat doing simple things, and then you will become an expert.

重复的事情用心做，就是赢家。

Do repeated things conscientiously, and then you will be rewarded with success.

所有成功的人生，

All successful lives

都是一个又一个梦想实现的叠加。

are a true follower of their dreams.

每一段不努力的时光，都是对生命的辜负。

Every time you do not work hard, you fail to live up to your life.

每日求知为智，内心丰盛为慧。

Wisdom comes from a thirst for knowledge every day and the richness of mind.

日有所学，月有所获，年有所成。

Learn day by day, and you will gain month by month and make achievements year by year.

当我们把学习、成长、精进和自我超越变成一种习惯时，

When we turn learning, growth, progress and self-transcendence into a habit,

我们的生命就会拥有一种向上向善的力量。

we will be supported by the power to be good and uplifting.

正是这种力量,

It is this power,

在无形中引领我们,进入多姿多彩的人生轨道。

that invisibly guides us to walk along an abundant life path accessible to merits and virtues.

所有成功的人生，
都是一个又一个梦想实现的叠加。

日积進

第 6 封信：

读书让你拥有智商

Letter 6: Study Offers You Intelligence

读书让你拥有智商，

Study offers you intelligence,

读人让你拥有情商。

and study gives you emotional intelligence.

生命之灯因热情而点燃,

The lamp of life is lit by enthusiasm,

生命之舟因拼搏而前行。

while the boat of life sails forward because of struggles.

坚持学习的人学到知识,

Those who stick to learning can acquire knowledge,

坚持练习的人学到本领。

while those who keep practicing can learn skills.

拥有梦想只是一种智力,

Having a dream is just a matter of intelligence,

实现梦想才是一种能力。

but making a dream come true is attributed to ability.

人生只有做好该做的事情,

Only when doing something due well in life

才有机会做我想做的事情。

can I have a chance to do what I want to do.

付出有多少，结果会说话。

The results are manifestations of the efforts.

过程会演戏，结果不骗人。

You may pretend in the process, but the result speaks for itself.

学习是最好的转运术。

Learning is the best way to turn your luck.

学习是身体与心灵的度假。

Learning is a holiday for your body and mind.

学习是最好的心灵美容。

Learning is the best beauty care for your soul.

书卷中得智慧，

Books can offer wisdom,

阅读是与古今圣贤相往来。

and reading makes us accessible to ancient and modern sages.

积极思考成就积极人生，

Positive thinking makes us rewarded with a positive life,

消极思考成就消极人生。

while negative thinking leads to a negative life.

危机对弱者来讲是灾难，

A crisis is a disaster for the weak,

但对强者来讲则是机会。

but an opportunity for the strong.

人生只有承担更大的责任，

Only by shouldering greater responsibilities in life,

才能有更大的成长和成功。

can we grow better and achieve greater success.

生命的目的不仅仅是成功，

The purpose of life is not only to be successful

生命的目的更是成长和分享。

but also to grow and share.

当我去照耀别人的时候，
我也被别人时刻照耀着。

日精進

日精進 青/少/年/双/语/版
Make Progress Day by Day

演說

成傑智慧心語
演說的最高境界
就說的體悟兼修
就是體悟兼高境界
就說服的
就是說服自己

第 7 封信：

把语言化为行动

Letter 7: Put Your Words into Action

把语言化为行动，

Put your words into action

把学习化为习惯。

and make study into a habit.

把学习变成生命的一部分。

Make study a part of life.

不为学习与成长付出的人,

Those who never make any effort in learning and growth

将会为失败付出更大的代价。

will pay a higher price for their failure.

教育改变一个人的思维模式,

Education changes one's mindset,

训练改变一个人的行为模式。

while training changes one's behavior pattern.

人生所有外在的成功和成就,

All the external success and achievements in life,

都是内在成长和成熟的体现。

are reflective of internal growth and maturity.

我人生中所走过的每一步路，

Every step of my life

未来都会精彩地呈现给这个世界。

will speak for itself in the future.

凡事要三思，

Think twice before acting,

但比三思更重要的，是三思而行。

but what counts more is to take actions after thinking twice.

行动是治愈恐惧的良药，

Action is the best solution for fear,

而犹豫和拖延，将不断滋养恐惧。

while hesitation and procrastination will continue to feed fear.

我们要向比自己优秀的人学习，

We should learn from those who do better than ourselves

我们不要和比自己差的人计较。

and we should not fuss about those who perform more poorly than ourselves.

当学习像空气一样无处不在时，

When learning becomes a habit,

它就会时刻滋养着我们的生命。

it will nourish our lives all the time.

凡事要三思，
但比三思更重要的，
是三思而行。

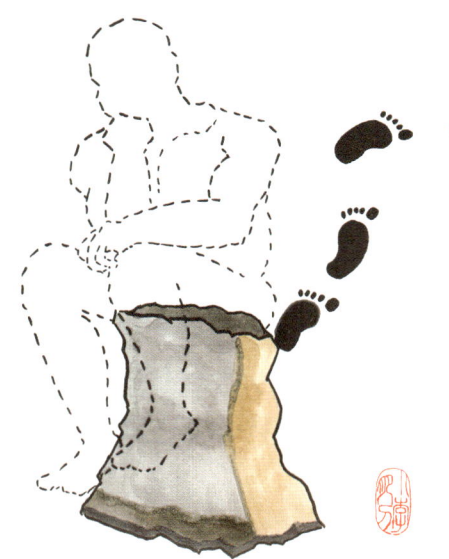

第 8 封信：

梦想的大小，决定成长的快慢

Letter 8: A Greater Dream Will Be Followed by a Quicker Chaser

梦想的大小，决定成长的快慢；

A greater dream will be followed by a quicker chaser.

格局的大小，决定成就的高低。

Great vision will be followed by great achievements.

人生最大的快乐，莫过于成长自己；

The greatest pleasure in life is nothing but having yourself become better.

人生最大的成就，莫过于成就众生。

The greatest achievement in life is nothing but helping those around you accomplish what they want.

人生成功的关键就是：找到对的人来合作。

The key to a successful life is to find the right person to work with.

学习成功的关键就是：向优秀的人学习和靠近。

The key to success in learning is to learn from and get close to great persons.

一个人只有成熟，他才懂得如何担当；

Only when one is mature can he know how to take on responsibilities .

日精进 青/少/年/双/语/版
Make Progress Day by Day

一个人只有担当，他才会更有魅力。

Only when one takes on responsibilities can he become more charming.

每天进步一点点，

Making a little progress every day

就是迈向卓越的开始。

makes the way for excellence start.

当我们把小事情做到极致的时候，

When we do small things perfectly,

老天爷就会给我们大机会。

God will award us a big opportunity.

生命中最值得怀念的日子，

The most memorable days in life,

就是那些最艰难的日子。

are those tough days.

当你战胜了困难,

When you overcome difficulties,

你就获得了成长和成功!

you will grow and succeed!

日積進 青/少/年/双/语/版
Make Progress Day by Day

中國夢

如实现中国梦无论何必须：以箱穹之洁求其苗，以泰岱之高求其基，以潭壑之深求其营，以天地之广求其反，气木共勉

覆天地不死
法古今完人

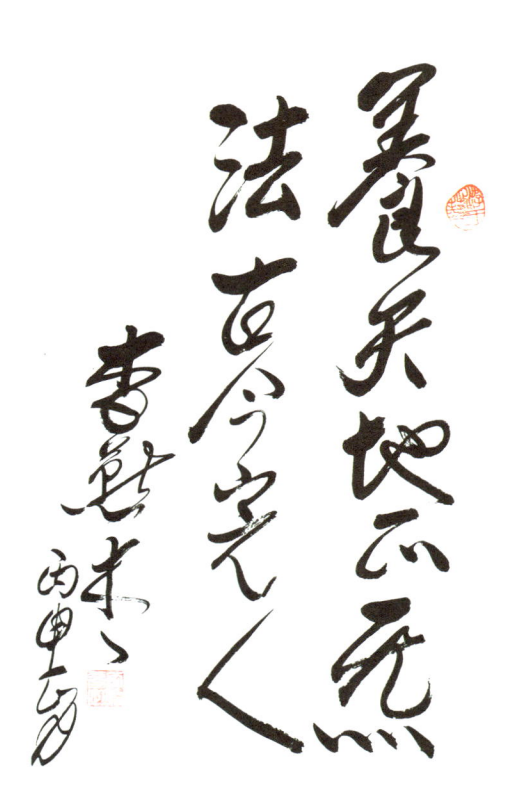

第 9 封信：

积极的人在忧患中看到机会

Letter 9: Positive People Find Opportunities in Hardship

积极的人在忧患中看到机会，

Positive people find opportunities in hardship,

消极的人在机会中看到忧患。

while negative people see hardship in opportunities.

路是脚踏出来的，历史是人写出来的。

One will reap what he sows, and history is created by people.

人的每一步行动，都在书写自己的历史。

People are writing their own history with their efforts.

人生只有做对的事，才会有好的结果；

Only by doing what is right in life can one be rewarded with good results.

人生只有做困难的事，才会有所收获。

Only by overcoming difficulties can one get himself improved.

一个有水平的人，怎么能允许自己不出成绩；

A talented man will not allow himself not to loaf through life

一个有水平的人，怎么能允许自己做得不好。
or perform badly.

聪明的人绝不等待机会，
A wise man never waits for an opportunity,
而是追求机会，掌控机会。
but pursues and proactively grasp opportunities.

什么是挫折？
What is frustration?
挫折其实就是：迈向成功所应缴的学费。
Frustration is actually what one must pay on the way to success.
什么是成功？
What is success?
成功就是走完了所有通向失败的路，
Success is the last way left,

只剩下最后一条，就通向成功。

after trying all the ways to failure.

成功是什么？
就是走过了所有通向失败的路，
只剩下一条成功的路。

第 10 封信：

遇见明天更好的自己

Letter 10: Meet a Better You Tomorrow

少而好学，如日出之阳；

Being young and studious, like the rising sun.

壮而好学，如日中之光；

Being postadolescent and studious, like the sunshine at noon.

老而好学，如炳烛之明。

Being old and studious, like the finite candlelight.

有一条路不能选择，那就是放弃；

The way that cannot be chosen is giving up.

有一条路不能拒绝，那就是成长。

The way that cannot be rejected is growth.

年轻时，我们播下什么样的种子；

One will reap

年老时，我们就收获什么样的果实。

what he sows.

先知三日，富贵十年。

One with foresight and insight can expect a bed of rose in life.

日积进 青/少/年/双/语/版

Make Progress Day by Day

今天我们所有的学习、成长、蜕变，

All our efforts in learning, growth and transformation today

都是为了遇见明天更好的自己。

are for making ourselves better tomorrow.

没有谁，富有得不要别人的帮助；

No one is too wealthy to seek others' help.

也没有谁，贫穷得一点也无法帮助别人。

or is too poor to help others at all.

要想拥有美好的人生，

To have a good life,

必须提升学识和能力，

a wealth of knowledge and abilities are necessary.

学识和能力的提升永无止境。

Learning much more knowledge and improving our abilities should be a long-term undertaking.

有知识才有见识,

Knowledge leads to insight,

有见识才有胆识。

and insight brings courage.

成功不属于最有条件、最有能力的人,

Success does not come to those who are the most qualified and capable,

成功属于最渴望、最相信、最愿意付出的人。

but those who desire, believe and work for it diligently.

年轻时，我们播下什么样的种子；
年老时，我们就收获什么样的果实。

没有谁，富有得不要别人的帮助，
也没有谁，
贫穷得一点也无法帮助别人。

第 11 封信：

成功是累积而成的

Letter 11: Success Comes from Constant Efforts

实现梦想并非人生的终点，
Realizing dreams does not mark the end of life,
实现梦想是人生新的起点。
but a new starting point in life.

把自己做好，就是对自己最大的贡献；

Being the best yourself is the greatest reward to yourself

把自己做好，就是对家庭最大的贡献。

as well as to your family.

在这个世界上，

In this world,

唯一可以不劳而获的就是贫穷；

poverty is the only thing that can be unearned without any labor.

在这个世界上，

In this world,

唯一可以无中生有的就是梦想。

dream is the only thing that can be out of thin air.

成功不是将来才有的，

Success does not come in the future,

而是从决定去做的那一刻起

it is realized by continuous efforts

持续累积而成的。

from the moment you decide to do it.

信心好比一粒种子，

Confidence, like a seed,

除非播种，否则不会结果。

will not bear fruit unless it is sown.

一旦目标确定，绝不瞻前顾后，

Once the goal is set, do not hesitate,

而要勇往直前，把1%的希望

just go forward bravely to turn 1% hope

变成100%的现实。

into 100% reality.

凡是不爱学习的人，

None of people who dislike learning

都不是真正追求进步的人；

will pursue progress from the bottom of their heart.

凡是不爱学习的人，

None of people who dislike learning

都不是真正热爱自己的人。

will really love themselves.

做好自己，即是爱与贡献；

Being yourself means both love and contribution.

做好当下，即是美好未来。

Doing well what you are doing now can lead to a bright future.

成杰智慧心语

爱

爱没有增加一切都是枉然
爱一旦增加一切即将改变

以古以立之公为
生民立命继
圣继绝学为
万世开太平
查越木
甲午秋

第 12 封信：

成为时代的弄潮儿

Letter 12: Being a Trendsetter of the Times

只有不畏攀登的采药者，
Only herb-diggers who are brave enough to climb
才能登上高峰采得仙草。
the peak can collect fairy grass.

只有不怕巨浪的弄潮儿，

Only seamen who are not afraid of huge waves

才能深入水底觅得珍宝。

can swim deep into the water to find treasures.

人与人之间最小的差距是智商，

What differs least among people is IQ,

最大的差距是坚持。

while what differs most is persistence.

成功就是站起来的次数，

Success means that the times you try

比倒下的次数多那么一次。

one more time than those you give up.

每一个人在奋斗中都会遇到困难、挫折和失败，

Everyone will encounter difficulties, setbacks and

failures in their life,

不同的思维和心态，是成功者和普通人的区别。

positive thinking and mentality can help one manage to overcome them and gain success.

每个人的人生只能活过一次，

Everyone can only live once,

而读书却可以让我们：

while we can learn about

体验千种人生，

thousands of lives,

感知千种活法。

and living styles in books.

我们最初拥有的只是梦想，

All we had at first are dreams

以及毫无根据的自信，

and baseless self-confidence,

但是，所有的一切，都从这里出发！

however, every hope comes from dreams and self-confidence!

每个人的人生
只能活过一次
而读书却可以让我们
体验千种人生
感知千种活法

第 13 封信:

命运并非机遇,而是一种选择

Letter 13: Destiny is Not an Opportunity, But a Choice

命运并非机遇,命运是一种选择。

Destiny is not an opportunity, but a choice.

我们不应该服从命运的安排,

We should not obey the fate,

我们应该凭借努力创造命运。
but create destiny with efforts.

人生中的每次付出，
Efforts in life are
就像山谷中的喊声，
like shouts in the valley,
你不需要谁能听到，
they don't need to be heard
但那延绵悠远的回音，
because the long-lasting echo
就是生活对你最好的回报。
is the best return of life for you.

那些尝试去做某事却总是失败的人，
Those who keep trying but always fail

日精進 青/少/年/双/语/版

Make Progress Day by Day

比那些什么也不尝试的人，

are infinitely better than

不知要好上多少倍。

those who try nothing.

伟人所达到并保持着的高度，

The heights reached and maintained by great men

并不是一飞就到的，

are not reached in a short time,

而是他们在别人偷懒的时候，

instead, it is credited to their efforts

一步步艰辛地向上攀爬而达到的。

in climbing up hard step by step when others are lazy.

这个世界并不是掌握在那些嘲笑者的手中，

The world is not held by those who mock it

而掌握在能够经受得住嘲笑与批评，不断往前走的

人手中。

but by those who can stand ridicule and criticism and keep moving forward.

这个世界并不是掌握在那些嘲笑者的手中,
而掌握在能够经受住嘲笑与批评,
不断往前走的人手中。

第 14 封信：

学习的敌人是自己的满足

Letter 14: Complacency is the Enemy of Study

精进的人，没有空虚；

Making progress can defeat emptiness.

奋斗的人，没有遗憾；

Diligence will leave no regrets.

舍得的人，没有痛苦；

Being generous will drive pains away.

放下的人，没有纠结。

Being relaxed will cause no entanglement.

渴望是拥有的开始，

Possession begins with desire,

越渴望就会越拥有，

the more you desire, the more you will gain,

渴望度决定行动力。

because desire drives action.

不面对怎么担当？不担当怎么成长？

How to take on responsibility if you don't face it? How can you improve yourself without taking on responsibility?

不成长怎么强大？不强大何来独立？

How can you become strong without growth? How can you become independent without being strong?

越分享，越喜悦；越分享，越自信；

The more you share, the more joy you will enjoy.

The more you share, the more confident you will be.

越分享，越富有；越分享，越丰盛；

The more you share, wealthier you will be.

The more you share, the more colorful life you will enjoy.

越分享，越绽放；越分享，越飞翔；

The more you share, the more you will grow.

The more you share, the higher you will reach.

越分享，越有功德；越分享，越有福报。

The more you share, the more merit you will have.

The more you share, the better you will be rewarded.

学习的敌人是自己的满足,

Complacency is the enemy of study,

要认真学习一点东西,必须从不自满开始。

so you must get rid of complacency to study seriously.

对自己,学而不厌;对他人,诲人不倦。

Never stop learning for yourself and never be tired to teach others.

我们应该采取这种态度。

This is the attitude that ought to be followed.

每一次发奋努力的背后,必有加倍的奖赏。

There must be double rewards behind every hard work.

日精進 青/少/年/双/语/版

Make Progress Day by Day

所谓成功的人，
就是能够
用别人向他投掷的砖块，
来为自己建造一个稳固的根基。

第 15 封信：

人生是自我期许的结果

Letter 15: Life is a Product of Self-expectation

人生是自我期许的结果，

Life is a product of self-expectation,

人生是自我要求的结果，

life is a product of self-demand,

日積進 青/少/年/双/语/版

Make Progress Day by Day

人生是自我超越的结果。

life is a product of self-transcendence.

成功的开端是制定目标，

The beginning of success is to set goals,

成功的关键是采取行动，

the key to success is taking action,

成功的条件是锻造自我。

the condition for success is forging oneself.

世上最重要的事，不在于我们身在何处，

What counts most in the world is not where we are

而在于我们朝着什么方向走。

but lies in which direction we are going.

忍别人所不能忍的痛，

After enduring pain that others can't endure

吃别人所不能吃的苦，

and suffer bitterness that others can't bear,

将会收获别人得不到的收获。

we will reap what others can't get.

一个人之所以成为重要的人，

The reason why a man becomes important

是因为这个人做了重要的事；

is because he has done something important.

一个人为什么有机会做重要的事？

Why does a man have the opportunity to do important things?

是因为这个人把每一件小事都做到极致。

It is because he does every little thing really well.

用人品去感动别人，用改变去影响别人。

Touch others with human character and influence others

with change.

用状态去燃烧别人，用实力去征服别人；

Influence others with state and conquer others with strength.

用行动去激励别人，用坚持去赢得别人。

Motivate others with actions and win the respect and appreciation of others with persistence.

世上最重要的事，
不在于我们身在何处。
而在于
我们朝着什么方向
走！

教育

戊傑智慧心評

教育的核心價值在於激發一個人的想象力和創造力
教育的終極目的在於塑造一個人的使命感和價值觀

第 16 封信：

教育的核心价值

Letter 16: The Core Value of Education

教育的核心价值在于：

The core value of education lies in

激发一个人的想象力和创造力；

to fire others' imagination and creativity.

日精進 青/少/年/双/语/版

Make Progress Day by Day

教育的终极目标在于：

The ultimate goal of education is

塑造一个人的使命感和价值观。

to help others build a sense of mission and values.

在人生的道路上，

In the path of life,

如果你没有耐心去等待成功的到来，

if you don't have the patience to wait for success,

那么，你只好用一生的耐心去面对失败。

you have to face failure with life-long patience.

人生学习的三个阶段：

There are three stages of learning in life,

向古圣先贤学习，

learn from ancient sages,

向宇宙万物学习，

learn from everything in the universe,

向内心深处学习。

and learn from the bottom of your heart.

人生的三大幸事：

There are three blessings in life,

年轻的时候，遇到好老师；

when you are young, you meet good teachers,

中年的时候，遇到好搭档；

when you are in your middle age, you meet good partners,

年老的时候，遇到好学生。

and when you are old, you meet good students.

人生有三种活法。

There are three styles of living.

活出来：追求的是人生的成功与梦想。

First, live a life in this way, the pursuit in life is success and dream.

活精彩：追求的是人生的价值与意义。

Second, live a wonderful life in this way, the pursuit is the value and significance of life.

活明白：追求的是人生的智慧与觉醒。

Third, live a conscious life in this way, the pursuit is the wisdom and awakening of life.

爱自己最好的方式；
就是成长自己；
爱众生最好的方式；
就是成就众生；

日精進 青/少/年/双/语/版

Make Progress Day by Day

成長

愛自己最好的方式
就是成長自己
愛眾生最好的方式
就是成就眾生

——盛傑智慧心語

第 17 封信：

每次失败都会使人更坚定

Letter 17: Every Failure Makes People Firmer

所谓成功的人，

The so-called successful men

就是能够用别人向他投掷的砖块，

are those who can build a solid foundation for themselves

日精进 青/少/年/双/语/版

Make Progress Day by Day

来为自己建造一个稳固的根基!

with bricks thrown at them by others!

每一次失败,都会使一个勇敢的人更坚定。

Every failure makes a brave man firmer.

如果没有失败的刺激,我们或许甘愿平庸。

Without the stimulus of failure, we might settle for mediocrity.

失败使人发奋图强,只有历经失败的痛苦,

Failure makes people work with stamina and diligence, and only through the pain of failure

才能找到真正的自我,感受到真正的力量。

can one find the true self with real power.

失败是什么?

What is failure?

没什么，只是更走近成功一步。

Nothing but one step closer to success.

成功是什么？

What is success?

就是走过了所有通向失败的路，

It is the last way left

只剩下一条成功的路。

after going through all the failure.

安全感不是别人给的。

The sense of security is not something to be given.

你若不成长、不精进、不改变，

Without any growth, improvement and changes,

谁又能给你真正的安全感呢？

who can make you feel safe?

我们这代人最大的发现就是:

The most important discovery of our generation is

人类可以通过改变自己内心的态度,

that one can change his life and his world

来改变自己的人生,改变自己的世界。

by changing his inner attitude.

机会是吸引来的，
当我足够好的时候，
我就会有强大的吸引力。

第 18 封信:

积极的人控制环境

Letter 18: Positive People Govern the Environment

消极的人受环境控制,

Those who are negative are governed by the environment,

积极的人却控制环境。

while those positive govern it.

环境不会十全十美，

The environment will not be perfect,

我却可以全力以赴。

but I can spare no pains.

用大编剧的思维，来谱写我的传奇；

Compose my legendary life with the mind of a great scriptwriter.

用大导演的视野，来布局我的人生；

Arrange my life with the vision of a great director.

用大演员的状态，来演绎我的精彩。

Perform my brilliant life with the state of a great actor.

对于意志坚定、永不服输的人来说，

A man with great determination and unyielding spirit

永远不会有失败。

will never be defeated.

他会跌倒了再爬起来,

He will roll with the punches every time,

即使其他人都已退缩和屈服,

even if everyone else has cowered and succumbed to failure,

他依旧永远不会放弃。

he still will never give up.

机会是吸引来的,

Opportunities are to be attracted,

当我足够好的时候,

when I'm good enough,

我就会有强大的吸引力。

I will be endowed with strong appeal.

成功的真谛：

The essence of success is that

无论你是谁，无论你从事何种工作，

whoever you are or whatever you do,

成功都包括以下四点：

success will not come before

明白人生的目的；

you know what life is for,

从事你擅长的工作；

do what you excel,

发挥你最大的潜力；

use what you've got,

播撒造福社会的种子。

and benefit our society.

日精進 青/少/年/双/语/版
Make Progress Day by Day

歸零

厚德載物

彭清一

目錄進

第 19 封信：

梦想是人生的导航仪

Letter 19: Dream, The Navigator of Life

梦想是人生的导航仪。

Dream is the navigator of life.

梦想是生命的发动机。

Dream is the engine of life.

梦想是生命飞翔的翅膀。

Dream is like the wings of life.

爱没有增加，一切都是枉然；

It is all in vain if there is no more love.

爱一旦增加，一切即将改变。

It will change everything if love grows.

梦想的三大核心：

To make a dream come true, one should make it clear,

我要成为什么样的人？

Who I want to be?

我要成就什么样的事业？

What I want to accomplish?

我要成就多少人的梦想？

How many dreams do I have to fulfill?

梦想的四大标准：

A dream can be well supported by

热血沸腾、不可思议、奋不顾身、不枉此生。

great enthusiasm, great ambition and indomitable, spirit in this way, one can make his life worth-living.

精神源于思想，

Spirit derives from thought,

思想源于信念，

while thought comes from faith,

信念源于经历和体验。

and life's events and experiences are at the root of faith.

知而不行，不为真知；行而不知，不为真行。

Knowledge needs to be tested by practice, and practice needs to be based on knowledge.

格局一旦放大，美好即将发生。
梦想一旦放大，成绩即将上升。
行动一旦放大，实力即将提升。

第 20 封信：

学习获得知识

Letter 20: Learn to Acquire Knowledge

学习获得知识，练习拥有本领；

Learn to acquire knowledge, and practice to possess skills.

体验进入核心，分享传承智慧。

Experience to get the core, and share to inherit wisdom.

只要用心,就有可能;

Nothing is impossible with attention.

只要开始,永远不晚。

It's never too late to start.

一心所向,无所不能;

One can accomplish anything with firm goal.

一心所向,无所不达。

One can go anywhere with solid target.

一个向着伟大目标奔跑的人,

For one who runs toward his great aim,

全世界都会给他让路。

the whole world will make way for him.

物质终将被遗忘,

Material will eventually sink into oblivion,

日精进 青/少/年/双/语/版

Make Progress Day by Day

唯有精神生生不息。

what remains eternal is our spirit.

日日精进,

Enhance yourself day by day

持之以恒。

with perseverance.

每日前进一步,

Taking one step every day,

日久可至千里。

one can walk a thousand li over the years.

知识改变命运,学习成就未来,

Knowledge changes the destiny, while learning shapes the future.

梦想指引方向,行动创造辉煌。

Dream gives the direction, while action makes glorious achievements.

第 21 封信：

格局一旦放大，美好即将发生

Letter 21: A Beautiful Life Comes from Wide Horizon

格局一旦放大，美好即将发生；

A beautiful life comes from wide horizon.

梦想一旦放大，成绩即将上升；

A better performance comes from great dreams.

行动一旦放大,实力即将提升。

A self-enhancement comes from amplified actions.

成功是梦想、格局、行动和坚持的高度配合。

Success is a good integration of dreams, horizons, actions and perseverance.

学习的态度决定成长的速度,

The speed of growth counts on the attitude of learning.

做人的态度决定成就的高度。

The height of achievement counts on the attitude of behaving.

学习不是学表面,而是学源头;

To learn the source rather than the surface.

开悟不是听结论,而是去探索。

To enlighten yourself by exploration instead of

passive acceptance.

热爱是最好的导师,

Passion is the greatest mentor

热爱是成功的源泉,

and the source of success

热爱可以跨越一切障碍。

which encourages people to surmount all obstacles.

多学习,善于思考;

Learn more and think better.

多行动,善于感悟;

Act more and be good at thinking.

多坚持,善于反省;

Be persistent and introspective.

多付出,方可收获。

No pains, no gains.

日精進 青/少/年/双/语/版

Make Progress Day by Day

物以类聚，人以群分。
当你身边出现高手的时候，
就是成长机会来临的时候。
与高手同行，
并成为高手。

第 22 封信：

与高手同行并成为高手

Letter 22: Become a Master-hand Under the Guidance of Master

物以类聚，人以群分。

As the saying goes, birds of a feather flock together.

当你身边出现高手的时候，

Whenever there is a master around you,

就是成长机会来临的时候。

it's time to learn from him.

与高手同行，并成为高手。

Become a master-hand under the guidance of master.

要在人前显贵，先在人后准备。

To be an outperformer in front of others, be prepared behind others.

成功来自充分的准备。

If you want to succeed, you must be fully prepared.

你不为成功做好准备，

Without preparation,

那就只能为失败所累。

you will doom to failure.

没有人能替你思考，

No one can think for you.

没有人能替你行动,

No one can take action for you.

没有人能代替你取得成功,

No one can take your place to make a success for you.

更没有人为你的失败买单。

No one can pay for your failure.

一切的一切,都靠你自己!

You are your own master!

普通人喜欢慢慢来,

Average person likes to take things slow,

高手选择立即行动。

while the master chooses to jump into action.

日積進 青/少/年/双/语/版
Make Progress Day by Day

学而知不足
不足而知学

成就智慧心语

戊戌年於文渊写
李俊平

一个学生最大的失败不是成绩差，
而是梦想毁灭，
信念坍塌，
放弃希望！

日精进

第 23 封信：

学习的最高境界

Letter 23: The Highest Realm of Learning

学习的最高境界是：

The highest realm of learning is

学以致用，马上行动。

to use what you learn to practice and take actions immediately.

一个学生最大的失败不是成绩差,

The biggest failure of a student is not poor academic performance,

而是梦想毁灭、信念坍塌,放弃希望。

but the decease of dream, belief and hope.

学习成功分两半:

Two causes work for success in learning,

一半在命运手中,那是天命;

one lies in destiny, that is god's will,

一半在自己手中,那是拼命。

another in your own hands, that is hard work.

人生是体验的盛宴,

Life is the feast of experience

生命是体验的总和。

and the sum of experience.

学习的最佳状态是：

The optimum state of learning is

建立强大自我，不受一切干扰。

to build a strong ego free from all interferences.

自信、自在、自然、自强、自如！

Be confident, comfortable, natural, strong, and free!

唯有巅峰的状态，

Only at the peak state

才能造就巅峰的成就。

can one make brilliant achievements.

随时保持优良状态，

Keep a good state at any moment

才能保证优美姿态。

to ensure your outcomes.

第 24 封信：

内心强大则世界辽阔

Letter 24: Stronger Heart, Wider World

内心变强，则世界辉煌；

A stronger heart can see a more splendid world.

心宽一尺，则路宽一丈。

A wider heart, a wider path.

日積進 青/少/年/双/语/版
Make Progress Day by Day

人生不如意十之八九。

Sometimes things just don't work out the way we thought they would.

我不能改变世界，

The world can't be changed,

但我可以改变自己。

what can be changed is myself.

心海宁静，宠辱不惊；

Be peaceful with tranquil mindset.

心胸宽广，无惧无畏。

Be fearless with open mind.

失败不必耿耿于怀，

You don't need to hold the grudge about failure.

人生需要激情满怀。

Life needs to be filled with passion.

我是我命运的主宰，

I am the master of m y destiny,

我是我灵魂的统帅，

captain of my soul,

我是我梦想的行者，

practitioner of my dream

我是我精神的导师，

and mentor of my spirit,

我活在当下，

I live in the moment,

我用心书写，

wholeheartedly,

我全力以赴，向上向善！

I will spare no pains in the pursuit of a good life!

智慧而深远，仁爱而持重，勇决而笃实，博识而谦恭

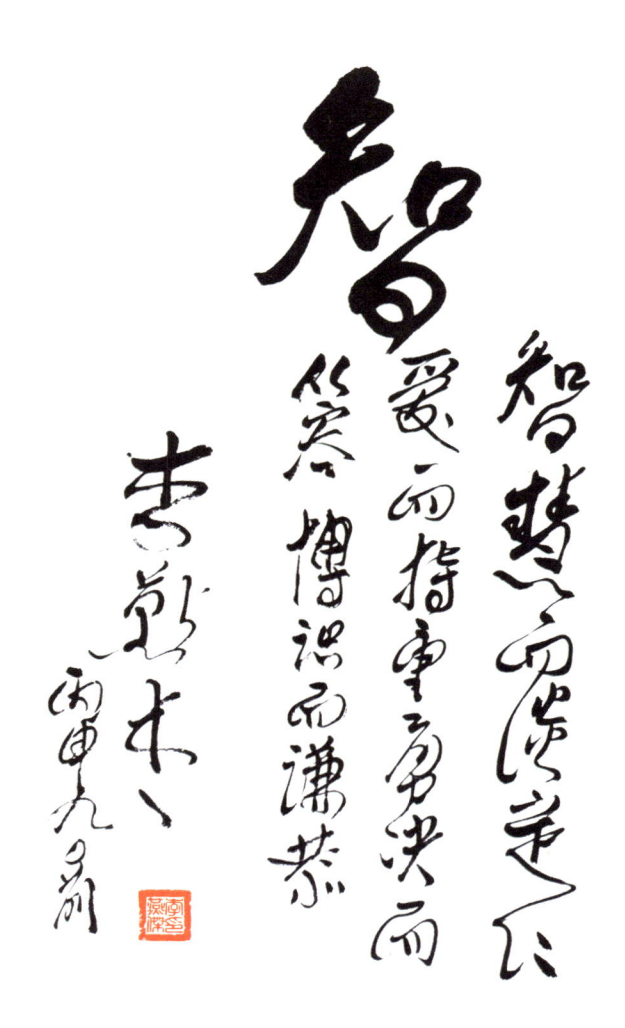

孝

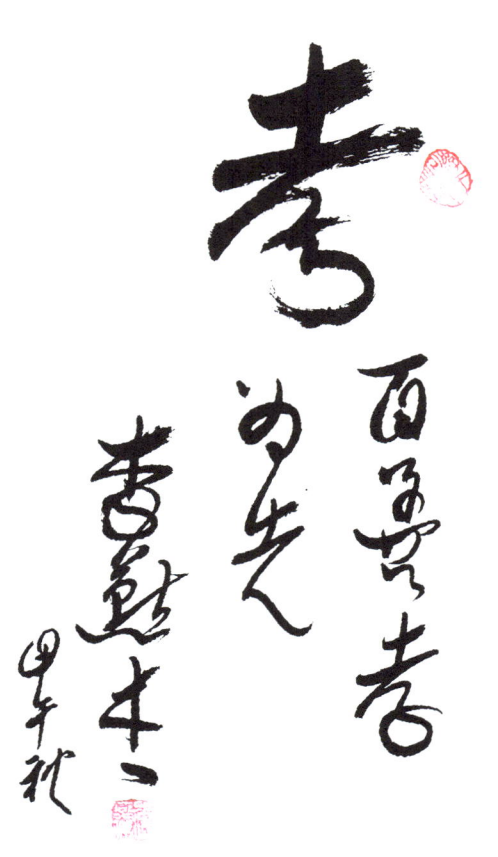

百善孝為先

甲午秋 日揆進書

第 25 封信：

没有经历付出就不会杰出

Letter 25: No Pains, No Gains

没有经历付出，就不会杰出；

No pains, no gains.

没有经历牺牲，就不会重生；

No sacrifice, no rebirth.

没有经历痛苦，就不会强大。

Adversity makes a man wise.

每时每刻，做有价值的事情；

Do something valuable at every moment.

争分夺秒，做有意义的事情。

Make full use of every minute to do something meaningful.

博学是一个人的学识与精进，

Erudition accompanies with one's knowledge and enhancement.

仁爱是一个人的仁慈与爱心。

Kindheartedness grows with one's kindness and love.

一个人能让自己兴奋起来，

To make yourself excited

是非常棒的事。

is something amazing.

一个人能让自己持续兴奋，

To keep this excitement

是很伟大的事。

is something great.

谦虚使人成长，精进让人飞翔；

Modesty makes one growing, enhancement soaring.

好学使人上进，坚持让人辉煌。

Studiousness makes one motivated, persistence brilliant.

你若能控制你的欲望，

Control your desire

你就能精通你的生命；

then master your life.

你若能坚守你的梦想，

Keep your dream

你就能掌控你的人生。

then control your life.

你能够承担多大的责任，你就能享受多大的荣耀。

第 26 封信：

目标一旦确定，方法就会出现

Letter 26: A Clear Goal is Followed by Ways

目标一旦确定，方法就会出现；

Once the goal is definite, there will be many ways,

目标一旦确定，动力就会无限；

infinite motivation,

目标一旦确定，实力就会展现；

great strength,

目标一旦确定，效果就会显现。

and noticeable outcomes.

水滴石穿，

Drops of water outwear the stone

并非水本身的力量。

not because the water itself

而是坚持的力量。

but the persistence.

绳锯木断，

Cut a wood with a rope

并非绳子本身的力量。

not because the rope itself

日精進 青/少/年/双/语/版
Make Progress Day by Day

而是信念的力量。

but the belief.

让我经手的每一件事，

Everything I did

都要贴上卓越的标签。

shall be labeled as excellence.

当我们把小事情做好时，

Do well in every little thing,

命运就会给我们大机会。

then destiny will offer a great chance.

人生只有跟随，才能获得精髓；

Follow your path to obtain the essence.

人生只有感恩，才会获得感动。

Show your gratitude to get moving moments.

一言之辩重于九鼎之宝
三寸之舌强于百万之师

第 27 封信：

开卷有益，开口有才

Letter 27: Be Beneficial from Books, Be Talented from Speaking

习惯于使用吉祥的语言，

With auspicious language,

我的人生就会吉祥如意;

everything goes well.

习惯于使用消极的语言,

With inauspicious language,

人生的光彩将逐渐黯淡。

the luster of life will gradually fade.

眼睛是心灵的窗户,

As the eyes are the window of the soul,

语言是智慧的门户。

language is the door of the wisdom.

优美的语言是才华的流露。

Beautiful language reveals one's talents,

奋进的语言是激情的流露。

enterprising language reveals one's passion.

台上一分钟,台下十年功。

One minute on the stage and ten years of practice off the stage.

养兵千日，用兵一时；

It takes a thousand days to train an army but one battle to use it.

千日造船，一日过江。

It takes a thousand days to build a boat but one day to cross the river.

一言之辩重于九鼎之宝，

A word of argument can weigh more than the treasure of nine tripods.

三寸之舌强于百万之师。

A silver tongue can be stronger than millions of soldiers.

只要讲话，就要真诚表达；

As long as you speak, it is wise to express yourself sincerely.

只要开口,就要学会赞美。

As long as you talk, it is wise to give praises.

人生有四乐:

One can enjoy the pleasures of life if he can

自得其乐,

enjoy himself,

知足常乐,

count his blessings,

助人为乐,

help others,

天天快乐。

and be happy every day.

人类的每一次进步,都离不开语言开路。

Every progress of mankind benefits from the charm of language.

日日进 青/少/年/双/语/版
Make Progress Day by Day

敬天爱人

正念利伢

第 28 封信：

志不立，天下无可成之事

Letter 28: No Vision, No Accomplishments

志不立，天下无可成之事。

Without a clear ambition, nothing can be accomplished.

德不立，世上无可靠之人。

Without noble moral trait, no one can be trusted.

做人，良知为本；

Having conscience is a basic requirement for every person.

做事，责任为本。

Taking on responsibility is the foundation of every career.

修己方能安人，内圣才能外王。

A gentleman cultivates himself and thereby brings peace and security to others, and a sage is an intellect who internally possesses virtue and outwardly acts as a ruler.

心态决定状态，眼界决定境界；

The mindset determines the state, and the realm of vision determines the prospects.

格局决定结局，能量决定力量。

The vision determines the outcome, and the energy determines the strength.

动机善，事必成；

Good motivation is followed by a success.

初心坏，事必败。

Bad original intention will be returned with a failure.

敬天爱人，正念利他；

A person will benefit from his behavior of respecting nature and others.

乐于助人，善于助人。

Therefore, we should be ready to help others at any time.

水洗万物而自清，人利众人而自成。

Water can clean everything while keeping itself pure, and a man can make accomplishments while helping others.

自私自利是小我，乐于助人是大我，

Selfishness signifies the ego, while helpfulness the

collective.

废寝忘食是忘我，敬天爱人是无我。

Forgetting to eat and sleep reflects immersion, and respecting nature and others reflects selflessness.

一个人从无到有是建立自我，

Establishing oneself from scratch

凭的是能力与拼搏；

depends on ability and hard work.

一个人从有到无是追求无我，

Pursuing selflessness in possession

修的是胸怀与境界。

shows one's great mind and vision.

日積進 青/少/年/双/语/版

Make Progress Day by Day

只要用心，
就有可能；
只要开始，
永远不晚。

利眾者偉業必成

一致性內外兼修

成傑智慧心語 戊戌桃月下澣聞霄書於舍得居

日精進

第 29 封信：

利众者伟业必成

Letter 29: Great Achievements are Made for the Public Good

利众者伟业必成,
Great achievements are made for the public good,
一致性内外兼修。
consistency calls for coordinated internal and external efforts.

做事精益求精，

Seek persistent improving in work,

做人追求卓越。

and pursue excellence in establishing oneself.

天下本无事，随时皆自在。

There is nothing in the world, so you can enjoy yourself at any time.

世上本无事，庸人自扰之。

Our worst misfortunes are those which never befalls us.

得之不喜，失之不忧；

Gains should be greeted with no joy, and losses with no sorrow.

得失随缘，心无增减。

Gains and losses are the matter of fate, but our moods will not change with it.

志不可满，傲不可长，

Never be complacent and arrogant,

欲不可纵，乐不可极。

and never be overly indulgent and epicurean.

竹密不妨流水过，

Dense bamboo forests do not hinder the flow of water,

山高岂碍白云飞。

and towering mountains do not stop the white clouds.

认识一个人，推开一扇门；

Knowing someone is like opening a door.

阅读一本书，找到一条路。

Reading a book can help us find a way.

听君一席话，胜读十年书。

One day with a great teacher is better than a thousand days of diligent study.

差一点失败,叫成功;

A near miss from failure is success.

差一点成功,叫失败。

A near miss from success is failure.

成功需要付出代价,

Success comes at a price,

放弃需要付出更大的代价。

but giving up requires more.

不要怕失败,不为成功设限。

Don't be afraid of failure, and don't set limits for success.

人生不设限,才能精彩无限。

Life can be wonderful without limits.

活出生命的精彩,

Live a wonderful life,

日精進 青/少/年/双/语/版

Make Progress Day by Day

我比我想象得更有力量。

I'm stronger than I thought.

不怕万人阻挡，
只怕自己投降。

第 30 封信：

地低成海，人低为王

Letter 30: The More Noble, the More Humble

地低成海，人低为王。

The more noble, the more humble.

小成功靠朋友，大成功靠对手；

Friends and hardship help one achieve small success,

小成功靠磨难，大成就靠灾难。

while opponents and tragedies help one accomplish great

success.

经历了磨难，就获得了提升；

Hardships help one get improved,

战胜了灾难，就获得了重生。

and triumph over tragedy helps one greet with a radical change.

心不唤物，物不至。

All wishes come true with the aid of expectation.

成功始于决心，

Success starts with determination,

成就源于渴望。

while achievement comes from desire.

渴望是拥有的开始，

Possession begins with desire,

越渴望就会越拥有。

so the more you desire, the more you will gain.

渴望的程度决定拥有的程度,

The more you desire, the more you will have,

渴望的速度决定成功的速度。

and the sooner you desire, the faster you will succeed.

付出有多少,结果会说话。

The results are manifestations of the efforts.

今天的收获,是过去的付出。

Today's gains are credited to past efforts.

明天的收获,是今天的努力。

Tomorrow's gains to today's efforts.

有付出才会走向杰出,

Only when you make efforts will you become outstanding.

有拼搏才会收获成果。

only when you work hard can you reap the rewards.

第 31 封信：

所有世间乐，悉从利他生

Letter 31: All Happiness in the World Comes from Helping Others

能理解别人对自己的不理解，就是胸怀；
Broad-minded people can understand those who do not understand them.
能包容别人对自己的不包容，就是大爱。
People with great love can tolerate those who

do not tolerate them.

一个付出的人，不会贫穷；

People who know how to give will not be poor.

一个索取的人，不会富有。

People who only take will not become rich.

所有世间乐，悉有利他生；

All happiness in the world comes from helping others.

一切世间苦，咸由自利成。

All suffering in the world is caused by self-interest.

自私自利让人变得渺小，

Selfish people will be insignificant,

放下私利，也就放下了渺小的自己；

so one will not be insignificant if he becomes not selfish.

日 精 進 青/少/年/双/语/版

Make Progress Day by Day

无私无我让人变得伟大，

Selfless people will be great,

选择无私，逐渐成就了伟大的自己。

so one will be great if he is selfless.

昨天已经过去，未来还未到来。

Yesterday has gone, and the future is not coming.

现在的一切，就是最好的安排。

Everything now is the best arrangement.

人生大约三万天，

There are about 30,000 days in a person's life,

要把一生过好的秘诀，

so the secret of living a good life

就是过好每一个今天。

is to take every day seriously.

仰望星空，脚踏实地。

Look up at the stars but come down to earth.

活在当下，活出精彩。

Live in the present and live a wonderful life.

人生要懂取舍，
人生要知进退。

日精進 青/少/年/双/语/版
Make Progress Day by Day

以心为师
智慧如海

飢心為師
智慧如後

成傑智慧心語

戊戌宏摩书

第 32 封信：

智慧而淡定，仁爱而从容

Letter 32: Wise and Composed, Kind and Calm

成大业者心中有爱，无恨；
The one bearing a great achievement has love in heart.
成大业者心中坚信，无疑；
The one bearing a great achievement has a firm faith.

成大业者心中宽容，无堵。

The one bearing a great achievement is tolerant.

智慧而淡定，仁爱而持重；

One should be wise and composed, kind and solemn,

勇决而从容，博识而谦恭。

brave and calm, erudite and humble.

立功立德立言真三不朽，

The one who makes contributions, cultivates virtue, and expounds ideas in writing is immortal,

明理明知明教乃万人师。

the one who is reasonable, knowledgeable, and good at teaching is the top teacher.

慈悲为本，利他为先，焦点利众，众人成全。

Take compassion as the foundation and put benefiting

others and the public first, and then they will help you make accomplishments.

品若梅花香在骨,人如秋水玉为神。
The character should be the same as plum blossoms, and the spirit the same as beautiful jade.
人的品德应该像梅花一样芳香入骨,
The character should be as noble and pure as plum blossoms,
人的精神应该像美玉一般晶莹剔透。
and the spirit as clean and honest as beautiful jade.

小成者,活在自我的世界;
The one who has achieved small achievements only focuses on themselves.
大成者,活在众人的世界。
The one who has achieved great achievements

cares about others.

学识渊博不是为了炫耀，而是可以唤醒他人；
Being knowledgeable is not for showing off, but for enlightening others.

财富丰厚不是为了享乐，而是可以帮助他人；
Being rich is not for pleasure, but for helping others.

地位显赫不是孤芳自赏，而是可以率众前行；
Being eminent is not for self-admiration, but for guiding others.

能量强大不是欺负弱小，而是可以赋能众生。
Being strong is not for bullying, but for delivering others from torment.

日精進 青/少/年/双/语/版

Make Progress Day by Day

学习是智慧的升华
分享是生命的伟大

彭清一题

读圣贤书行仁义事
立修齐志存忠孝心

第 33 封信:

不辩,是一种智慧

Letter 33: It is a Kind of Wisdom not to Argue

不辩,是一种智慧;

It is a kind of wisdom not to argue.

不争,是一种慈悲;

It is a kind of compassion not to quarrel.

不看，是一种自在；

It is a kind of ease not to look.

不闻，是一种清净。

It is a kind of sobriety not to listen.

立身中正，为事公允；

Be impartial and fair.

大其心，容天下之物；

Be tolerant and accept all things of the world.

虚其心，受天下之善；

Be humble and accept the benevolence of the world.

平其心，论天下之事；

Be calm and discuss the affairs of the world.

潜其心，观天下之理；

Be concentrated and learn the principles of the world.

定其心，应天下之变。

Be quiet and adapt to the changes in the world.

从现在看过去，会看见无知；

Looking back, you will find ignorance.

从宽容看是非，会看见解脱；

Tolerate right and wrong, and you will be free.

从接受看命运，会看见踏实；

Accept destiny, and you will be quiet.

从平凡看生活，会看见喜悦；

Enjoy ordinary life, and you will be happy.

从反思看内心，会看见成长；

Reflect on yourself, and you will grow.

从乐观看未来，会看见希望；

Be optimistic about the future, and you will see hope.

从知足看人生，会看见珍惜；

Be content with life, and you will cherish everything.

从反省看自己，会看见转机；

Introspect yourself, and you will find a favorable turn.

从随缘看世界,会看见自在。

Follow the fate, and you will be at ease.

广种福田,广结善缘,

Do more kind deeds, make good connections,

广交朋友,广行大道。

make more friends, and be moral.

第 34 封信:

人生的成长是日精进和随时学

Letter 34: Make Progress Every Day and Learn at Any Time for Improvement

人生的成长是日精进和随时学,

Make progress every day and learn at any time for improvement.

人生的幸福是身亦安和心亦宽，

Be healthy and patient for much more happiness,

人生的实践是管住嘴和迈开腿。

and control what you eat and do appropriate exercise for a better life.

有梦想的人，不会懒惰；

The one who has dreams will not be lazy.

有格局的人，不会计较；

The one who is tolerant will not be calculating.

有境界的人，不会纠结。

The one who has good mental state will not struggle for something.

我要做自己的主人，

I will be my own master,

日精进 青/少/年/双/语/版

Make Progress Day by Day

我要做生命的主人,

I want to be the master of life,

我要和我在一起。

and I want to follow my heart.

我和我在一起,是一次久违的重逢;

For a long-lost reunion,

我和我在一起,是一种身心灵的合一;

for the harmony of body and mind.

我和我在一起,是一份期盼许久的向往;

for a long-awaited yearning.

我和我在一起,是生命自然的觉察;

for being conscious about the nature and life.

我和我在一起,是生命真实的存在与显现。

and for the true existence and manifestation of life.

拥有爱的人是快乐的,

People who are loved are happy,

付出爱的人是幸福的，

who love others are happy,

充满爱的人是温暖的。

and who are with love in their hearts are kind-hearted.

用爱人的心做事，用感恩的心做人！

Love others, and be grateful to others forever!

日精進 青/少/年/双/语/版

Make Progress Day by Day

智慧

每日求知為智
内心豐盛為慧

成傑智慧心語

戊戌孟宪章

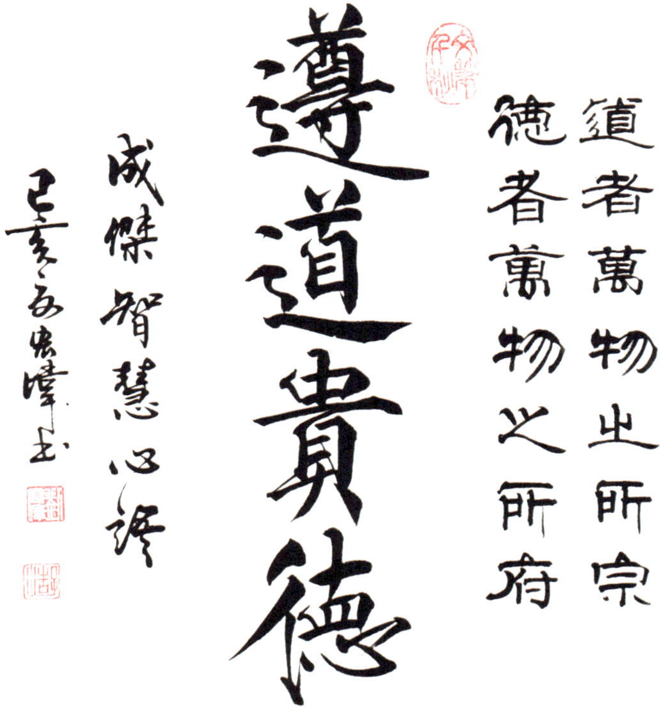

道者萬物之所宗
德者萬物之所府

遵道貴德

成傑 賀慧 心語

日積進

第 35 封信：

平常心是道，道是平常心

Letter 35: An Ordinary Mind is Excellent Practice, and Excellent Practice Comes from an Ordinary Mind

平常心是道，道是平常心。

An ordinary mind is excellent practice, and excellent practice comes from an ordinary mind.

道不外求，以心为师。

Excellent practice cannot be learned from others, it should be taught by our mind.

万法由心生，万法由心灭。

The emergence and disappearance of everything are determined by you.

万术不如一道，万法不如一心。

All skills are not as good as excellent practice and all methods are not as good as one's mind.

心乃众智之要，

Mind is the key to all wisdom,

心是一切智慧的源泉。

and the source of all wisdom.

天地不可一日无和气，

Heaven and earth cannot exist for a day without harmony,

人心不可一日无喜神。

and people cannot live for a day without happiness and strength.

心量决定能量，
Mind determines energy,
能量决定力量。
and energy determines strength.

只要用心，就有可能；
Nothing is impossible with attention.
只要开始，永远不晚。
It's never too late to start.

对学习，怀一颗进取心；
Study hard,
对生活，怀一颗感恩心；
be grateful to life,
对生命，怀一颗敬畏心；

be reverent for life,

对父母,怀一颗孝敬心;

be filial to parents,

对人生,怀一颗欢喜心。

and love life.

心宽一尺,路宽一丈。

A wider heart, a wider path.

心若计较,处处都是怨言;

If you are calculating, you will complain everywhere.

心若放宽,时时都是春天。

If you are tolerant, you will be happy everyday.

心就是一切,一切源于心。

Your heart is everything, and everything comes from your heart.

日精進 青/少/年/双/语/版

Make Progress Day by Day

生命智慧的十大法門

生命的擁有在於時時感恩
生命的能量在於焦點利眾
生命的偉大在於心中有夢
生命的強大在於歷經苦難
生命的喜悅在於傳道分享
生命的價值在於普度眾生
生命的綻放在於內在豐盛
生命的幸福在於用心經營
生命的成長在於日日精進
生命的蛻變在於真正決定

摘自成傑智慧心語 荊霄鵬書

巨海成杰希望小学视频

第 36 封信：

生命智慧的十大法门

Letter 36: Ten Ways of Life Wisdom

生命的拥有在于时时感恩，

Always being grateful makes people rich in spirit.

生命的能量在于焦点利众，

Incessant helping others gives people great energy.

生命的伟大在于心中有梦，

Great dreams make life out the ordinary.

生命的强大在于历经苦难，

Painful experiences make people strong.

生命的喜悦在于传道分享，

Preaching and sharing bring happiness.

生命的价值在于普度众生，

The greatest value of life is helping the sentient beings.

生命的绽放在于内在丰盛，

The inner abundance makes life bloom.

生命的幸福在于用心经营，

Human's blessedness comes from cherish and manage life with great care.

生命的成长在于日日精进，

Step forward day by day will finally lead huge leaps.

生命的蜕变在于真正决定。

Life changes radically when people determined.

日粹進 青/少/年/双/语/版

Make Progress Day by Day

天行健 君子以自强不息

彭清

地勢坤　君子以厚德載物

彭清

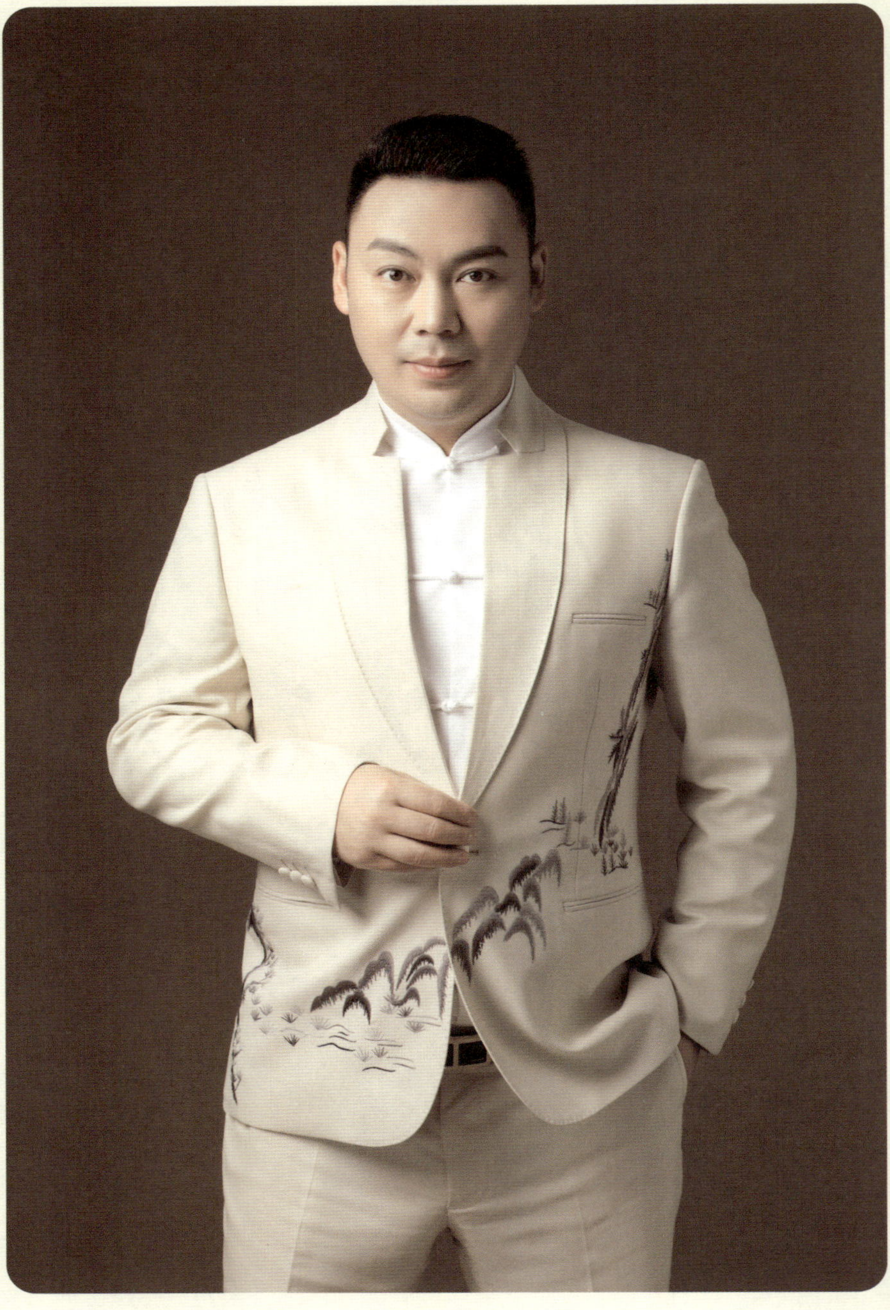

成 杰
CHENGJIE

巨海集团董事长
上海巨海成杰公益基金会发起人

成杰个人宣传片

他，立身教育。
以帮助中国民营企业发展为己任，在过去18年里走过165座城市，巡回演讲5600余场。帮助近千名企业家学会公众演说，直接听众达近百万人次。

他，为梦而行。
2008年创办上海巨海企业管理顾问有限公司，现业务已经遍布全国。

他，研学创新。
独创了"生命智慧的十大法门"，自主研发了"商业真经""为爱成交·国际研讨会""领袖经营智慧""演说智慧·终极班""大道之学""商界演说家"等一系列精品课程，并持续地更新迭代、完善升级。

他，著书立传。
出版了《日精进·道心卷》《日精进·初心卷》《日精进·明心卷》《掌控演说》《大智慧》《商业真经》《觉醒》《为爱成交》《80后演说少帅》《从优秀到卓越》《一语定乾坤》等畅销书。

他，助力公益。
立志用毕生的时间和精力来捐建101所希望小学。2015年，发起上海巨海成杰公益基金会，目前已成功捐建18所巨海希望小学。

他坐而论道，起而践行，不忘初心，坚守使命。以教育培训为终身事业，助力中国中小型企业发展与腾飞。

巨海集团介绍

上海巨海企业管理顾问有限公司是由成杰老师创办于2008年10月。

巨海集团宣传片

巨海集团是一家集巨海智慧书院、巨海商业学堂、未来领袖商学、企业实战管理、领袖魅力演说、企业内训、顾问式咨询诊断、演说家论道、领袖论坛为一体的专业咨询机构。巨海集团以"帮助企业成长，成就同仁梦想，为中国成为世界第一经济强国而努力奋斗"的使命为己任，立志成为中国商业培训优选服务平台！

巨海集团聚焦企业发展，研究各行业新、精、尖的企业经营管理资讯，整合优秀人才和市场优质资源。采用新的商业模式，帮助企业全方位成长，协助企业进行更有效的管理，提高全员职业素养，打造职业化团队，从而提升企业核心竞争力。

巨龙腾飞，海纳百川。巨海，是一个聚焦天下英才的舞台；巨海，是一个创造奇迹的事业平台；巨海，是一个拥有伟大使命感与崇高愿景的快速成长型企业。我们始终致力于：为同仁搭建极具成长性的事业平台；为客户提供极具实战、实用、实效的管理培训。

上海巨海成杰公益基金会

把爱传出去
生命更精彩

上海巨海成杰公益基金会　　　巨海希望小学